SHELBY IS ALIVE

CHUONG VAN NGUYEN

Copyright © 2022 by Chuong Van Nguyen

Paperback: 978-1-958381-30-4
eBook: 978-1-958381-31-1
Library of Congress Control Number: 2022911695

All rights reserved. No part of this publication may be reproduced, distributed, or transmitted in any form or by any electronic or mechanical means, without the prior written permission of the publisher, except in the case of brief quotations embodied in critical reviews and certain other noncommercial uses permitted by copyright law.

This is a work of nonfiction.

A bunch of young adults, three lads name Ian, Dan, Ham and a beautiful blonde babe name Shelby went out to the ford's dealership shop wanting to steal one of the brand-new Fords Mustang models.

When they are there looking around the shop full of variety of new sports cars, they found one that they like is the mustang GT FN model with a nice paint job done on it, with a very nice shiny red color.

They find that one to be enticing.

The four had a good final look at the car and walked off out of the dealership.

They agree to rob the store with that one car they had chosen but Shelby is having 2nd thoughts about it. She thinks she is the only girl in the group and that it will slow them down.

It doesn't matter, Ian said!

They left and went home that day and waited for three days for everything to go quiet and time passes by so nobody would take extra notice about them in an obvious way.

Three days have passed by, they all began dressing in black clothing during that night, they drove to that car dealership shop with equipment to unlock the doors from the shop and during that night they sneak into the back entrance opening the door with a crowbar. As soon as they got in, they were searching for that main car that they saw days ago because it is now put in somewhere at some place. Moments of searching, they have finally found that car. The car was hidden in the end corner where they couldn't see.

They all came to it, and later Ian tells Shelby to go find the keys to it. He also tells Dan to go with her!

Moments of searching the two have finally found the set of keys to all the cars hidden where the salesman have put it in the lock. When they break open it with the hammer, there was just too many keys in the suitcase, so it took them a while to go on a search for the right ones by clicking the alarm buttons of each one. While they both trying to search for the right key Shelby turns her head to the side standing looking at the room where they have not been to yet with colorful lights can be seen under the closed door. She walks to it while Dan is continuing looking for the key.

She opens the door and sees a so much better looking car then the one they wanted to steal. The car is electric supercharge Mustang Shelby GT500. She loved it so much because her name is too also Shelby, and her favorite color is blue and white on a car. She comes to it feeling and touching the beautiful vehicle. She notices the car is a prototype. The car can be seen connected by circuits and wires all over the car and to the machine, which looks to be unfinish project.

She opens the door from the car and sits in it. She loves the car so much and the scent of interior. When she was sitting there, she was fidgeting around with the buttons and accidentally press a few with her fingers because she heard Dan from a distance saying he found the right key to the car that they were going to steal which jump scare her for a sec. The car she is sitting went full alarm where the sirens wailing loud, the screen of the front displayed as danger mode. The doors suddenly closed on its own and locked Shelby inside where she could not escape herself.

The screen now displays as malfunctioned then the word danger over repeatedly. She cries for help! She tries to use force herself to break free by kicking the door window but there was no damage to it.

Dan in the other hand tries to rush in to see what was going on in the room and sees Shelby is stuck inside the vehicle. He sees her panicking, and the other two Ian and Ham came too to see what is going on.

She can be seen crying for help, so the three took out their equipment and try to break the car window and unlocking the door with a crowbar but everything went failure. The vehicle is unbreakable.

Ham tries to whack the wires and cables that is attached to the front and the back cable from the car off to see if the doors will unlock but made matters even worse. She can be seen from inside being electrocuted and smokes coming out of the sides of the doors.

There is nothing else the three could do about it, they were panicking themselves because they were afraid the cops might show up. For that moment they left her, but Dan tried his best wanting to stay a bit longer, but Ian and Ham pulled him and left the place.

She is though being electrocuted, but she could still see them. In her mind she said she will get them for this, for leaving her behind.

She screams very loudly in anger and pain from them leaving her behind and also from the electrocuted. She suddenly exploded and all the blood and body parts can be seen everywhere from inside. She died that very moment.

During that day in the morning, the sales team had arrived at the ford's dealership. When they came, they notice the back entrance was broken into. They came in and saw so many damages to their merchandise. One of the staff then tries to dial for 911 call for emergency.

The dealers wondering why there was no cops that came during last night, they must have broken in the alarms, say's one of the staff!

The boss of the ford's dealership told the staffs not to call the force because nothing is stolen but just a few damages to the properties.

One of the staff told the boss to come and look at the car prototype that they were working on. Someone has tampered with it inside and died. They all came and see blood stain everywhere in the car windows of the vehicle. The boss unlocks the car with the key that he took home for that one. There was just so much blood and body parts everywhere from inside. They all identify that person died in their vehicle is a young blonde woman. Later that day, they look at their surveillance camera to see what was happening during last night. They see everything and the boss wants to cover up everything, like as if nothing has ever occurred at their dealership at all.

About their prototype vehicle, they have decided to continue the project and sell it off to the sports club as they keep recommending to them.

Three months have passed, the prototype car is completed and is now sold to the sports club for the raffle winner to claim the prize.

Weeks of collecting the tickets from patrons, the game begins in the evening Saturday. The show down then begins that day. The raffle ticket machine now scrolls away. And the winner goes to Miguel.

He was so excited that he won the prize and went up to claim it. The spokesman asks him how he feels about winning the car and he said he couldn't be much happier because he has never won anything before in his life.

He was very nervous and shaky at first because the prize he won is not cheap.

Moments of celebration, he left afterwards with the car and went home during that evening.

When he came home, he told his family about his winning! They congratulate him on his new car. While they're standing there on the porch looking at the car talking, they all went back inside the house but except for Miguel, he was sitting there starring at his car for hours during the night with the lights on. He kept on mumbling saying that he can't believe that he won, repeatedly so many times.

During that night he went to his room laying on the bed couldn't sleep but helped it, he went out of his house again and sat on his new car while his other old car that his gramps has gave him is just sitting there in one spot being neglected.

He was feeling the interiors and loved the smell of the inside.

Out of nowhere his little sister's name Susan came out of the house also to see him and his new car. He asks her what she is doing outside at first and she told him that he only wanted to see his new car of his. So, she told him to come sit inside the car beside him.

They were talking and talking for a while until for that very moment she asks him what does most of these buttons do? She points one and another while he describes to her when she deliberately presses all the buttons at once with her fingers and laugh, the car started malfunctioning. The screen displays as danger mode, but it also unlocked a voice talking machine to the car. The female voice says danger mode, danger mode!

He turns off his car with the ignition key and everything stopped. He turns it back on his car and everything went back to normal as it seems. He yelled his sister in a soft voice to go back inside the house. She wanted to apologize to him, but he said it was alright and insist on her going inside the house. He felt agitated.

Moments of silence for a bit the car suddenly speaks. The car speaks to him out of nowhere through the radio station.

Her first word was, Hello!

He got kind of confuse thinking that it was some casting in the radio which was talking to each other but not him. He tries to turn off the radio station but its not seem to be working because its already off.

She asks him can he help her, and then he asks her "who is this" she replied to him saying this is Shelby.

She said she is stuck in this car and doesn't know how to get out. He got kind of confused, don't know what she is on about. He thinks she is one of those future upgrade car models where we could just talk to it.

None of this didn't happen before when I was driving it home, said he! My sister must have awakened this car by pressing multiple of buttons.

He asks her what's her name and she replied "Shelby"

She doesn't want him to be afraid as she said, because she is afraid herself! She doesn't remember how she got here stuck inside this car like as if its her body. She is wondering where her real human body is. She is so confused herself and has an amnesia.

Just for that split second her memory kind of pop-up bits by bits. She started to remember she was inside this mustang and somehow her body got exploded that she must have fused her soul into this car.

Her precious human body is now gone and that she is now trapped into this car for permanent. There is nothing she can do for now but to calm herself down.

She explains to him deep further about the problem that occurred on that day where she died. The more she talks to him the more she remembers. She remembers it was her male friends that left her for dead. She now remembers very clearly and wanted to avenge her death.

This is getting silly, he says.

He ignores her for the time being and told her that he needs to get to work tomorrow delivering pizzas. He said bye to Shelby and went back inside the house. She was upset that he just left her, but she stayed quiet mode for now.

The next morning around eleven he showed up to the pizza shop with his other old car. He doesn't want to take the new car along and show it out to his co-workers because he might damage it by driving too much and hit something that also scratches it.

Later that day, he finishes his shift and brings home some free pizzas for his family to eat. It was mainly supreme toppings. He gave the rest of the pizzas to them but took a few slices and went out to the front house to see his mustang. While he was standing at the porch eating the slice he went to the car and sat inside. He is sitting there eating when Shelby says Hi to him out of nowhere. He got a jump scare at first because he thought he was dreaming that night, he thought she was not real. He keeps on wondering why a radio can talk and respond to him like that. She tries to talk to him telling him that she is real, and she died in this car of his, but he is just being too stubborn to think about it. He always thinks she is one of those new car models where they can just talk to the passengers. She tries to talk to him, but he kept on neglecting her. He tried turning off the radio station, but it does not work, and it only angers her even more. You're only making me angrier; she says!

While he continues to ignore her, he finishes his slice of pizza and took the car for a drive. She kept on talking to him wanting him to believe her story is all true, but he is always refusing to believe her that she got fused and transferred her soul into the car. Until that sudden snap she got very upset, the car speeds off very fast and then it stopped in an instant. She tells and yells at him that if he does not believe her, then he should look it up on google and find the incident and the date that occurred on the day months ago at the dealership.

He finally listens to her about that one and searches it up. And when he does, he stops the car for a sec and looked it up, he could not find the news about her death in a car incident anywhere on the web.

That's a lie she says!

He said to her if he is lying then she can read through his smart phone by connecting the charger cable to the car and phone. She doesn't know if she can do that but told him to give it a try anyways! So, he did, and it was a success. It worked and she was able to search and scan the date of the incident, but she too unfortunately couldn't find the news anywhere on the web. It must have been a cover up she says!

Once again, he does not believe in her. So, she describes to him that she is a blonde beauty which most people say she is with a slim figure and was trapped in this mustang GT500 where she got electrocuted and has fused her soul into this vehicle permanently.

No matter what or how many times she explains to him, he will always keep on being stubborn no matter what, thinking she is just one of those new models where there are special features where cars are now receiving upgrades.

She is so very furious where she could not take it anymore then suddenly the car ignition turns on and she steps on the accelerator by herself but not him. The car was doing a burn out.

She continues to curse on him while the car was doing a burn out and he was so scared telling her to stop! He said he is sorry and now believes in her. So, she stopped, and the car suddenly went off at that instant and so as the burn out stopped as well.

Everything went quiet and he was panicking for a minute. he was calming himself down and said what was all that about. He knew what was going on but because he was scared that's why he would say that.

seconds later he asks her is she there, thinking what has happened because everything was a ruckus and suddenly shutdown in a quiet mode. Shelby did not respond back at him. So, he started up the engine peacefully and drove back home.

Later, he has gotten back to his house sitting there in his car quietly turning off the ignition, he wonders if he should tell his family about the car. Shelby then talks to him, and he suddenly got a jump scare from her. She tells him not to tell his family about her talking to him and that she is alive. She wanted everything to be kept a secret.

What would happen if I do tell someone about you said, he!

Then I will make the worse out of you, and here's an example said Shelby!

She turns on the alarm from the car and was wailing loud noise.

Is that a threat he says!

He went out to go find some sort of object and break the car. She said go ahead and give it his best shot!

He took a shovel and whack the car around so many times that the shovel end up breaking and did no damages to the car. he was like what the heck! What is this car made of?

This car is made almost bullet proof tough as steel, she says. There is no way you can destroy me.

Oh yeah, he said, he was going to take fuel from his garage and burn the car out. She yells at him and tries to threaten to run over him by revving the car at him. He said OK fine.

She can even make things even more worser then that to him if he keeps it up like that.

How is it that you're able to control the car most of the time, he asks?

She doesn't know either, but she might of guess it must be of anger or frustration that had tap and triggered her to control it. While she was explaining to him, his mom and dad plus Susan came out of the house standing at the front porch. His dad called for him asking if he was alright because of those car revving noises that they heard from inside.

Shelby whisper to Miguel, telling him not to say anything to his family about her or she will kill them and including him. So, he kept his word and only told his farther that he is alright and that its nothing major going on out here. They all went back inside the house including him as well.

The next morning Miguel was called by his employer telling him that his shift was off. So later that day, he heads off to the gym with his car Shelby. When he was driving, he tries to speak to Shelby but there was no reply from her. So, he kept on driving along without saying anything else to her and as soon he had arrived at the GYM, he turns off his car at the parking lot of the place and when he does Shelby scares him. She says boo to him!

Jesus, you scared me for a moment, says him!

After that she wanted to ask him for a favor. What's that he replies!

She wants him to take her to see her friends after when he finishes up with his workout. He said no, because he knows exactly what she is going to do next when she sees them.

She said she won't harm them because it's a different friend not the ones at the robbery dealership incident.

Besides that, he said, didn't you wanted not to let anyone know about your present state.

Of course, but only to them not anyone else.

He wanted to know what she is plan on doing when she meets up with them. She will let him know later after when he finishes his workout.

Afterwards he walked to the GYM and behind him from a distance were a bunch of guys ready to workout also. They walk pass and saw his car and was kind of jealous about it. They were looking at it for a while and then walked to the GYM.

While Miguel was working out, the guys from earlier that saw his car was also working out close by to him. They came up to him looking for trouble asking him how he got the car. they both were arguing such on and gotten into a fight and the manager of the facility came out trying to stop it. The manager was telling all of them off about fighting each other. The trouble making boys left the GYM and went out to the parking lot to their cars until they came up with the idea of wrecking Miguel's car while he is still talking to the manager inside.

One of the guys grabbed a baseball bat from his trunk and went to his car and bashed it and the guy was like what the heck! How come its not damaging his car! then the other guy uses his key to scratch his car and his paint left scrapped. The body cannot be damage, but the paint can. Shelby can be seen furious and couldn't take it anymore by them and responded back by turning on the car and running them over. They got ran over but they had minor damages to their bodies. They all ran off in fear.

Shelby then drives back to the same parking spot and stayed put.

After when Miguel finishes talking to the manager of the GYM, he heads outside to the parking lot to go home and when he does, he sees his car filled with scratches and marks everywhere. What the hell happen to my car, said him!

He felt very upset to see this has happen to his car. He knew it was them boys from earlier that tried to start trouble against him. He went sat in his car for a while until Shelby came on and spoke to him, she told him it was the boys who mess with you. He figures!

She told him don't worry about them because she just tries to run them over and pretty much scared the living out of them. They won't be back, she says!

Then she asks him for a favor about before earlier seeing her friends. Yes, he said, and he asks her again what she is going to do if she sees them. I'm not going to do nothing she replies, because this is a different friend, not the ones at the time of the robbery.

She misses her other friends and only wanted to see them for the last time.

In her mind she is hiding the truth. Her wrath is beyond imagination, and she is lying to him to get to her old friends that left her on that day. She couldn't stop thinking about it. She badly wants revenge but hiding the truth to him knowing that he won't help her to get to them.

He will do this favor for her because she did it for him, she scared off the troublemakers.

He said he will take her to see her different buddies but needed a rest first, feeling exhausted from all his work outs. So, he decided to take her for tomorrow instead. And she agreed!

The next morning, he took the car Shelby to see her friends that she mentioned yesterday. While he was driving, she was showing the directions how to get to them. minutes of driving they had almost gotten to the place. They finally arrived with a few blocks away from them standing at their houses. They both see them three. Shelby is angry just seeing them brings her memories. Miguel responds to her asking is that her other friends not the ones during the robbery incident, and she said no that's them three Ian, dan and Ham.

What, he says! You lied to me this whole time for me to get to them.

She said it was the only way for him to take her to see them. she wants him to kill all three by running them over.

No, he says!

She kept on telling him to do it, but he kept on refusing to her, so she got very furious and angry that she drove the car herself. She sped off very fast doing a burn out and chasing after them. She can be heard screaming along the way and the three Ian,

Dan and Ham turns their head to the side seeing the car doing a burn out and yelling.

And Dan was like is that the car that Shelby got stuck inside months ago! The three notices it and said holy cow that is that car and they started running away as fast as they can.

Shelby yells and cursing at them while she is chasing after them, saying they will all pay for this, for leaving me behind!

Miguel tells her to stop this madness, but she is all out of control and won't listen to him. So, he tried to stop it himself by applying on the handbrakes and stepping on the brake pads, but nothing is working at all. She is in control of the car now. The three run as fast as they could but they couldn't outrun her.

Ham was unfortunately the first one to be killed due to his overweight size. She ran over him back and forward a few times.

And 2ndly Ian got killed by her, chasing him along the way and crashing him onto the brick wall of the building. Blood can be seen spilling out of his mouth.

And for last, Dan ran and ran as he could try to escape her but ran into a dead end. He ran onto the top of the cliff edge of the road. It was a checkmate and he got no where else to run. He was cornered. He was standing there being so afraid and telling her begging for forgiveness. Miguel can be seen inside the car telling her to stop all this while she still got the chance.

She was reveling the car so badly wanted to knock him over the cliff. Dan explains to her that he tried his best to stay back and help her that day but the other two dragged him out of there and couldn't do anything else afterwards. Shelby yells to him saying he could have tried harder and runs over him knocking him out and fall all the way down to the ground and died on the spot, when Miguel told her not to. He can be heard screaming and cursing inside the car. He is very disappointed in her for not listening to him. He should have known all this was a lie from the start just so she could get revenge on them. he then tells her to unlock the doors for him to get out, and so she did.

While he got out and is walking, she wanted to know where is he going? He said he is going home, and he is leaving her behind. She drove along aside next to him asking him if he would want a drive home. She said she is happy now and would do anything for him. He's response was a no; he does not want a lift home.

And just for a second, she remembered she could not control the car unless if she was either angry or upset. She now learns how to tap into her emotions for her to control it. Besides that, he told her to bugger off some where else because he doesn't need her anymore because she just murdered some people.

Where would I go, said she! All the paperwork is under your name.

He said he would sell her to a new owner or to the wreckers.

Is that a threat, she says!

It's not a threat, he says! He only wants her out of his life.

If he decides to leave her, she'll makes matters even worse!

Think about the family, she says!

He just thought about that and didn't realize it for that moment, and it made him even hate her more just for bringing that up. For the sudden in his mind, he came up with the idea of how he can terminate her from existence. He kept the strategy to himself without not letting anyone knowing about it.

So afterwards he pretends to forgive her and wanting a ride home. Moments later while he was inside Shelby giving a lift home, he came up with another brilliant idea to get rid of her when he saw the racing game on the board of side highway. He asks her if she wants to go racing in a few days because he says she is good at chasing her old friends without hitting any angles.

If she could win the race for him, he says, and get one million dollars prize money then he promises to her that he needs to fix her up first with a new paint job to enter the race. She does that and he will keep her forever as he promises.

He crosses his fingers behind his back making a lie to her, but she does not see it. She was happier then ever to hear that from him and agreed with him. Moments later when he has arrived home and parked Shelby on the side of the house. He went inside his house to his room just laying there thinking many ways to get rid of the car. then he looked at one of his chemistry books to how he can make explosives. The next day he was on to it and him perfection it in a few hours. And when he also left his car to the paint factory for a respray. He spent all his life savings which he was supposedly to fix and re-paint his old car that was given to him for his birthday but did it all on Shelby. He felt bad for it spending all his money on her, but he hopes she could win the race to get the million which is worth a lot more than his savings.

The days are up, and the car is done and ready for racing day begins. He brings the newly freshly painted Shelby GT500 to the racecourse ready to race. Everything's packed and ready to go, Miguel is in his car lining up with the rest of the others. Shelby is very confident that she will win this race. The race then begins. Miguel is pretending to hold the steering wheel making fake turns along the way while she is doing all the work.

She does beautiful driving and turns without hitting any curbs or objects along the way.

They were the first ones on the lead and suddenly won the race seconds later.

Yes! Miguel says! He is happy to have won the race.

Everybody cheered for him. So, he got out of his car and went on the stage to claim his winning prize. He celebrated during that day. After that he went back to his car and ask Shelby that he should take over the driving for her now and left the racecourse. She asks him where he is going and where abouts is he taking us.

You'll see, he says!

Shelby kept on talking and talking with full of excitement that he won the money for him, and he pretended to talk and be happy to her in the conversation.

They have now arrived at their destination. Miguel took out his winning cheque and out of the car. he too left his bag filled with explosives inside of Shelby, but she doesn't know what's inside or anything about it.

He walked to the edge of the cliff where Dan died asking her does she remember this.

And yes, she does remember and what about it, she says.

He turns around looking at Shelby and reaches his hand to his pocket and clicked the button from the home-made remote control and the explosives went off from inside the car. the car went into flames bursting the interior leathers. The car windows are broken.

Not so tough and unbreakable after all huh, said him! You can die after all!

He was jumping with joy thinking that she is dead from the explosive that he made.

The engine machine went turning back on and Shelby can be hear screaming to him (I'll get you for this!)

Uh oh, said he!

She drives full maximum speed towards him, and he was cornered exactly like Dan was and got no where else to run.

He taunts her telling her to (come on!)

And at that split second of almost hitting him, he jumped to the side and there goes Shelby that flew off the cliff and hit the ground. She exploded even more then before.

YES, he says. He finally destroyed her. He finally killed Shelby.

He was so happy that he's strategy worked, and he no longer must deal with her anymore.

He then takes out his cheque money kissing and dancing around with it, until suddenly the cops arrived. They came to him but at first Miguel thought that he was being arrested. The cops wanted to discuss about his car that had killed and murdered a few people. They told him that he is not being arrested because they have proof and witnesses that can proven him not guilty of the murderers, so he is not being charged. The cops have now charged the dealership for the cover up and the death of Shelby and has enough evidence to dismantle the car hoping it won't come back alive again.

Thank God he said, because he thought he was getting arrested for that moment. They ask him if he would want a lift home by their police car and he said yes most definitely.

When he has arrived home, he went and hugged his entire family. He was very happy to see them because he almost died from the car. His family asks them where the new car and he told them it is destroyed and going to the wreckers. A month later his old car given by his gramps has fully been restored and freshly painted and was loved by him so much and he too also had moved his family into a new house to a different suburban location to live.

To be continue .....

www.ingramcontent.com/pod-product-compliance
Ingram Content Group UK Ltd.
Pitfield, Milton Keynes, MK11 3LW, UK
UKHW061139180426
11947UKWH00001B/2